"Wake up, Zollie!" the alarm clock called. Zollie opened her eyes as her robot dog Comet walked through the door carrying her clothes.

"Wash up and change," Comet said. "It is time to eat!"

Zollie climbed out of bed. Another day on Zors had begun.

In the kitchen a robot was busy with measuring cups and spoons. Three arms put plates full of food before Zollie and her mother and father. Other robots were busy too.

When they first moved here, Zollie liked watching the house cook and clean itself. But today she could only think about her friends on Earth.

School was no different today than it had been for several weeks. The teacher showed pictures of Zortian mountains. She asked questions that the children answered on their cell pads.

Some kids sent video-mail to each other. But no one sent a v-mail to Zollie.

At recess they put on special suits to play outside. Many children played basketball. The hoops were as tall as a two story building. It was easy to jump high because everything weighed less on Zors. Zollie wanted to play, but both teams had enough players.

Zollie was too shy to ask if she could join the girls jumping rope. So she played alone.

Comet met her at the door after school. He brought her a small page.

"You have mail," he told her.

"I am sure it is from Suzu!" Zollie slid the page into the mail reader. At once she saw Suzu on the screen and heard her laughing and talking.

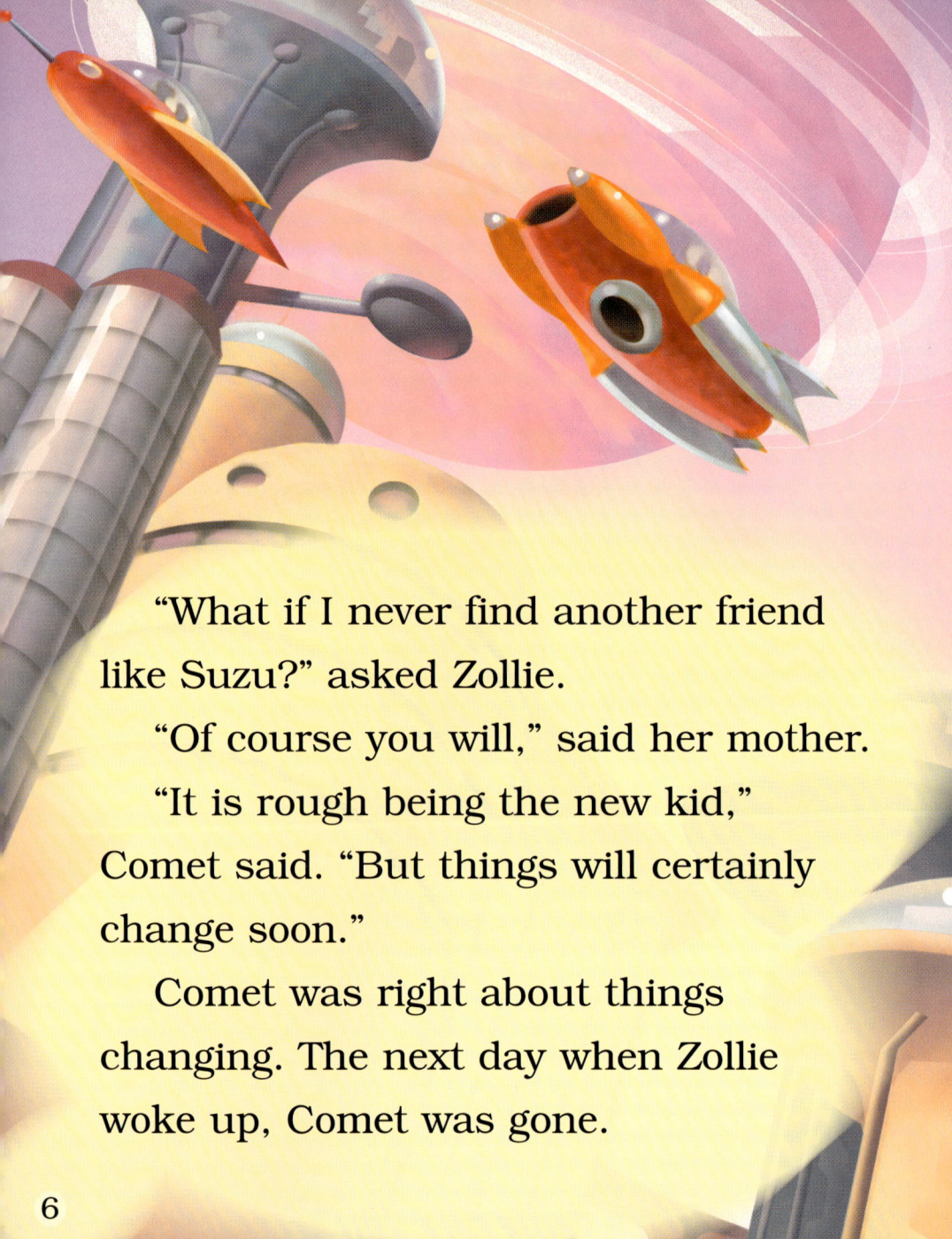

"What if I never find another friend like Suzu?" asked Zollie.

"Of course you will," said her mother.

"It is rough being the new kid," Comet said. "But things will certainly change soon."

Comet was right about things changing. The next day when Zollie woke up, Comet was gone.

Comet had never run away before. Zollie looked everywhere, calling him again and again. She thought it was strange that a clever robot dog like Comet would become lost.

Finally, Zollie heard sounds outside. She followed them and found Comet with a stranger.

"Comet!"

"I guess he's your dog," said the girl. "I am Nikky. I just moved here. I don't know if your dog found me, or I found him."

But Zollie knew. Comet had brought her a friend!